Collectible California Raisins™
An Unauthorized Guide with Values

Pamela Duvall Curran
George W. Curran

D0851684

4880 Lower Valley Road, Atglen, PA 19310 USA

ISBN: 0-7643-0433-X
Printed in United States

Published by Schiffer Publishing Ltd.
4880 Lower Valley Road
Atglen, PA 19310
Phone: (610) 593-1777; Fax: (610) 593-2002
E-mail: schifferbk@aol.com
Please write for a free catalog.
This book may be purchased from the publisher.
Please include $3.95 for shipping.
Try your bookstore first.

We are interested in hearing from authors
with book ideas on related subjects.

Contents

Acknowledgments

Aram Azadian, Sr.
Don & April Barcus
Tonita "Toni" Crittenden
Alice Curran
Larry DeAngelo
Aaron Matthew Duvall
Robert & Lois Duvall
Wendy Duvall
Allen Golomski
Stan & Dayle Golomski
David, Melanie, & Tyler Goodwin
Delores "Dee" Lawson
Karen Lynn Schaffer
Lynda Marie Schaffer
Peter Schiffer
Barry & Sue Ullmann
Paul Ullmann
Art Voorhees

It's *raisinable* to say that this book might never have been written if it weren't for some *grape* friends who are very serious *California Raisin™* lovers.

Toni Crittenden and Art Voorhees are friends and fellow collectors who have loved *The California Raisins™* since they were first introduced. Toni collected a lot of the early raisin products and material, sent away for many of the free commercial offers, joined *The California Raisins™* Fan Club, and purchased most of the fan club items offered to members. When we first mentioned our interest in raisins, we should have known that Toni would have a huge bag of goodies to drag out to the living room and show us. After seeing Toni's collection, we knew that our raisin hunts would intensify—*grapely*! Thank you so much, Toni and Art, for sharing your collection and information with all of us, and for sharing your goodies with George and I. You two are true collectors and we'll always cherish your special friendship.

When we contacted Barry Ullmann, we were thrilled to be invited to see his fabulous raisin collection and photograph some great pieces for this book. Barry and his son, Paul, spent long hours working with us; comparing their items with those already pictured for the book, setting up groups of related items for photographs, and carrying armloads of "raisin stuff" from one location to another. Thank you, Barry, Sue, and Paul, for your hospitality. Heartfelt thanks to Barry, from us and collectors everywhere, for sharing your outstanding collection.

As always, our family deserves our warmest thanks. Very special thanks to Melanie, Lynda, and Karen—you've all made some exceptional finds for our collection. *Raisin'* kids to hunt raisins was certainly worth the effort! To our family members who also enjoy collecting, we say thank you for the many wonderful hours we've spent together in pursuit of the assorted and elusive treasures we all love. Those are special times we'll hold in our hearts forever.

Thank you, once again, to Peter Schiffer. It is a privilege to work with such a professional and dedicated organization as Schiffer Publishing, Ltd.

Introduction

Television commercials introduced the dancing, singing *California Raisins*™ to the American public in the fall of 1986. Those sun-wrinkled Raisins were fresh from the sunny fields of Fresno, wearing sneakers, white gloves, and an occasional pair of sunglasses. The animated *California Raisins*™ characters quickly became a national phenomenon and their popularity soared beyond all *grape expectations.*

What started all this? The California raisin industry was looking for new and exciting ways to promote the wholesome goodness of raisins, and subsequently the California Raisin Advisory Board (CALRAB) was formed. Collecting assessments from the California grape growers and packers, CALRAB then utilized the funds to formulate an elaborate nationwide advertising and promotional campaign.

The original *California Raisins*™ Claymation® characters were created by Will Vinton, the acclaimed Claymation® designer. The Claymation® raisin commercials were voted by television audiences as the best TV ad campaign for the years 1987 and 1988. The *California Raisins*™ commercials even topped such giant advertising spenders as McDonald's, General Motors, and Pepsi-Cola. At this time, there are eight known Claymation® commercials, created and produced by Will Vinton Productions, that were played for American audiences: "Late Show," "Lunch Box," "Playing with Your Food," "Raisin Ray," "Locker Room," "Intermission," "Michael Raisin," and "Manager."

The leading manufacturer and distributor of licensed CALRAB raisin merchandise was Applause® Licensing Inc. of Woodland Hills, California. Most items will be marked, not only with the name CALRAB, but with Applause as the licensee, and the country of manufacture. Applause contracted with more than twenty-five leading companies to produce literally dozens of items, from the PVC plastics to T-shirts, books, games, and so on. Applause then marketed the items nationally and internationally through retail gift shops, card shops, and chain stores.

In addition to Raisins for the television commercials, Will Vinton Productions designed a whole cast of raisin characters for a number of Claymation® movies. Two of these—*The California Raisins*™ *Meet The Raisins* and *The California Raisins*™ *II, Raisins: Sold Out!*—were eventually released on videocassette. Collectors who are able to acquire and view these movies will become better acquainted with the names and roles of the many raisin characters.

The *California Raisin*™ mania progressed from television commercials and movies, to records and cassettes, a television cartoon show, a fan club and newsletter, and personal *California Raisins*™ appearances all over the country. Before long, *The California Raisins*™ went international, appearing in countries such as Canada, the United Kingdom, Australia, and Japan.

Grocery stores participated in in-store, on-package merchandising promotions. Products by Sun-Maid, Del Monte, Nabisco, and General Foods' Post® Natural Raisin Bran were among those utilizing *The California Raisins*™ characters to entice buyers. Encouraging repeat sales, some of these companies offered free raisin merchandise to consumers who sent in specified numbers of proofs of purchase.

The *Hardee's* Restaurant fast food chain also played a major role in promoting *The California Raisins*™ characters. Hardee's came out with four different series of raisins, starting in 1987, each available for a limited time. This promotion of raisins was the largest and most successful one that the restaurants had ever participated in. Hardee's are located primarily in eastern and southern regions of the United States, and their raisin collectibles are currently plentiful in those areas.

For a short period of time, *The California Raisins*™ had their own newsletter called *The California Raisins*™ *Grapevine Gazette,* which was mailed to Fan Club members. The first issue of the *Gazette* announced the results of a national contest to name *The California Raisins*™. In 1987, it was determined that the raisin characters would be more personal and lovable if they had names. A huge "Name The Raisins" contest was promoted in magazines such as *People* and *TV Guide*, and the Sunday newspaper comics. The winning names of Tiny Goodbite, Ben Indasun, and Justin X. Grape were chosen from more than 316,000 entries. The three winners who submitted the names won $5,000 each and a trip to Hollywood and Fresno, California.

The *Grapevine Gazette* also provided fans with up-to-date information on the antics of the raisin characters (real people dressed in raisin costumes) who traveled all over the country in a Winnebago doing promotional and guest appearances. There seemed to be no limit to where they went and what they became involved in. The Raisins visited the White House during the administrations of Ronald Reagan and George Bush, attending the Annual Christmas Tree Lighting Ceremony in Washington, D.C. They attended a private children's Christmas Party hosted by First Lady Nancy Reagan, and boogied their way down Pennsylvania Avenue for the Inaugural Parade of President Bush. Raisins danced and sang their way along the route of the Macy's Thanksgiving Day Parade in New York City for several years running. They promoted physical fitness programs in schools through the Great Raisin' Fitness Challenge; and joined with the American Library Association to instigate a reading program called "Books. Check 'Em Out!", which aimed to make reading fun and exciting for elementary school students. Both the fitness and reading programs rewarded students who participated and excelled in the programs.

Abruptly, the curtain came down, and the singing and dancing came to an end. Prompted by a petition of a majority of raisin packers, the State of California terminated the marketing program of the California Raisin Advisory Board in April of 1994, and CALRAB closed down in July of that year. Funds in their account were to be used to pay expenses related to closing, and then disbursements of the remaining funds were to be made to the California raisin growers and packers.

Authors' Note: It didn't take long for the popularity of the CALRAB raisin merchandise to generate an influx of cheaper, unlicensed imitations in the marketplace. All merchandise which was licensed and copyrighted by CALRAB will be marked as CALRAB. In general, these are preferred by collectors.

For most of you, the decision to collect *The California Raisins™* will begin with finding the figural PVC characters. As your collection of PVC's grows, you will note some minor discrepancies in facial expressions, eye color, eyelid color, hand positions, and the overall decorating quality. This is not unusual, and collectors will often justify having duplicates of an item, due to these differences.

We hope this book helps in *raisin'* your awareness of what is available, and that it will be a guide you carry with you at all times, and enjoy many times over.

Condition and Value

Price guides are the single most-scrutinized factor of any new collectibles book, and certainly can generate the most controversy. The prices in this guide should bring some sense of stability to the field of collecting *California Raisins™*, but above all else, should be considered as a guideline only. Many variables can determine what price is paid for collectibles: collector interest, scarcity, condition, and even geographical location. Collectors should remember that they make the final decision as to what an item is worth to them. No price guide can do that for you.

Throughout the book, references are made to Mint in Package (MIP), Mint in Box (MIB), and Mint on Card (MOC). These items will always be worth more to a collector than items that have been used or played with. In turn, items with considerable wear and tear will be worth far less to a collector than the listed price. Prices should always be determined based on a fair assessment of condition.

Ceramic or glass items should be in perfect condition to command listed prices. Chips, hairlines, crazing, stains, discoloration of graphics, etc., will decrease values. Mechanical, video, musical, and other items with working parts, should all be in excellent condition to command listed prices.

Chapter 1

Figural Characters

The hard plastic PVC figural characters in this first chapter were copyrighted by CALRAB; licensed, manufactured, and distributed by Applause™ Licensing, Inc.; and some will show copyrights by Will Vinton Productions. All were sold in retail gift shops, card stores, or chain stores; or were offered as promotional giveaways by such companies as General Foods and Del Monte.

Often played with by children then tossed into toy boxes, it should be noted that prices listed are for those items in very good condition, with no paint wear or damage. Each figure is marked with a copyright date, CALRAB, Applause™, and country of manufacture. Some of the later characters are also marked "Claymation® © Will Vinton."

Some PVC's have been given a variety of unofficial nicknames by collectors, primarily for identification purposes. Only the listed names/descriptions given to the characters by CALRAB/Applause/Will Vinton Productions appear in this book.

Dimensions, as always, are approximate. Figures were measured to the top of the head, since measuring to the highest hand position would create too many inconsistencies.

Ben Indasun, 2.5" original conga dancer with orange sunglasses and sneakers, 1987, made in China. Value $5-7.

Tiny Goodbite, 2.5" original microphone singer, 1987, made in China. This version has the longer microphone held in the left hand, unattached to the body. Value $8-10.

Justin X. Grape, 2.5" original conga dancer with blue sneakers, 1987, made in China. Value $5-7.

Saxophone Player, 2.5" original sax player, 1987, made in China. Value $5-7.

Winking Man, 2.5" with hot pink sneakers, 1988, made in China. First edition released in January 1988. Value $10-12.

Red Guitar Player, 2.5" with turquoise sneakers, 1988, made in China. First edition released in January 1988. Value $10-12.

Microphone Singer, 2.5" with a shorter microphone in the left hand and molded to the face, 1987, made in China. First edition released in January 1988. Value $5-7.

Santa Hat raisin man, 3" with green sneakers, 1988, made in China. **Candy Cane** raisin man, 2.5" with red sneakers and green laces, green sunglasses, 1988, made in China, released May 1988. Value $10-12 each.

Turquoise Sunglasses raisin man, 2.5" with turquoise sneakers and sunglasses, 1988, made in China. This version has sunglasses which have shorter sides and appear to be molded to the face. Right hand points up, and left hand points down. First edition released in January 1988. Value $10-12.

This side view shows the two previous Turquoise Sunglasses raisin men. Open eyes can be seen under the sunglasses of the man on the right.

Turquoise Sunglasses raisin man, 2.5" with turquoise sneakers and sunglasses, 1988, made in China. This version has sunglasses which appear to be added to the original molded head, giving a view of his open eyes under the glasses, and both hands point forward. First edition released in January 1988. Value $14-16.

Valentine Man, 2.5" with red sneakers, 1988, made in China. **Valentine Woman**, 2.75" with hot pink shoes, 1988, made in China, released October 1988. Value $10-12 each.

Drummer, 3" raisin man, black hat with yellow feather, 1988, made in China. Second edition released in August 1988. Value $18-20.

Bass Player, 2.75" raisin man with blue and gray boots, 1988, made in China. The Bass Player on left is shown in plastic package, which includes an application to join *The California Raisins*™ Fan Club. Second edition released in August 1988. Value MIP $18-20; $15 loose.
Note: Four raisins, *Bass Player, Drummer, Pink Shoes Female, Yellow Shoes Female*, were offered on the *Post Raisin Bran* cereal boxes for free by sending in six proofs of purchase or two proofs of purchase plus $3.99 for the set.

Pink Shoes Female, 3" with pink shoes and bracelet, no tambourine, 1988, made in China. Second edition released in August 1988. Value $15.

Ms. Marvelous, 2.75" with green shoes and bracelet, tambourine, sideswept hairdo. While the other two females are marked on the soles of their shoes, this piece is marked on her backside, 1988, made in China. Offered through the newsletter for $1.50 postage paid to fan club members. Value $15.

Yellow Shoes Female, 2.75" with yellow shoes and bracelets, tambourine, 1988, made in China. Second edition released in August 1988. Value $15.

Horizontal Surfboard raisin, 2.5" high, made in China. Value $50-60.

Horizontal Surfboard pictured from back, showing that surfboard is not attached to sneaker.

Michael Jackson Raisin, 3-3/8" high, was a special release in September 1989. This moonwalking, silver-gloved raisin stands on his own base, which is marked with the following: The California Raisins™ © CALRAB © Will Vinton Prod. Inc. © Triumph Lic. by Applause Lic. China. Value $20-25.

The Hip Band series, made in China. Third edition released in December 1988. Left to right: **Hip Guitarist**, 2.5", yellow guitar, pink headband. Value $18-22. **Boy with Microphone** in right hand, 2.25". Value $15-18. **Girl with Microphone**, 2.5", yellow shoes. Value $15-18. **Sax Player**, 2.5", wearing black beret. Value $15-18.

Vertical Surfboard raisin, 2.5" high at head, 1988, made in China. Value $35-40.

Raisin' Some Fun in the Sun series, made in China, released in March 1989. Value $12-15 each. Left to right: **Girl with Sunglasses**, 1.75", sitting on sand with radio. **Boy with Surfboard**, 2.5", standing on sand. **Girl with Grass Skirt**, 2.75", not on sand base like the others. **Boy in Beach Chair**, 1.5", sitting in striped chair on sand, with bottle of suntan lotion.

Vertical Surfboard pictured from back, showing surfboard attached to sneaker.

Meet the Raisins series, released May 1989, all marked: Claymation®, ©Will Vinton, Applause™, China. Left to right: **Red the Piano Player**, 2.5" high at top of head. Value $30. **Lick Broccoli**, 3" high, plays a guitar. Value $18-20. **Rudy Bagaman**, 3" high, holding cigar. Value $18-20. **Banana White**, 3" high. Value $18-20.

Red the Piano Player, 2.5", is shown mint in package, and without package. Value $40 MIP; $30 loose.

Meet the Raisins II series, made in China, released September 1989. Left to right: **Cecil Thyme**, 3-3/8" high, wears gray pants and sport coat. Value $110-125. **A.C.**, 3-1/8" high, wears red sneakers and extends his left hand. Value $110-125. **Leonard Limabean**, 2.5" high, wears a lavender suit and blue hat, and sports a handlebar mustache. Value $110-125. **Mom (Lulu Arborman)**, 2.5" high, wears pink apron and blue shoes. She's the mother of A.C. and Beebop. Value $110-125.

The Graduates are the original four raisins with a mortarboard added, each is approximately 3.5" high, including the yellow plastic base. The base has a sticker which is marked: Applause™ The California Raisins™ © 1988 CALRAB Claymations designed by Will Vinton, Made in China. Left to right: **Justin X. Grape** graduate, blue sneakers, base reads: "Heard it through the grapevine, you graduated. Congratulations." **Ben Indasun** graduate, orange sunglasses, base reads: "You really danced through it .... Congratulations!" **Tiny Goodbite** graduate, microphone singer, base reads: "Time to sing your praises. Congratulations Grad." **Sax Player** graduate, base reads: "World's most hell raisin' grad!" Value $40-50 each.

Box for the Del Monte Fruit Snacks musical sandwich. (*Courtesy of Toni Crittenden & Art Voorhees*)

Close-up of the sandwich PVC's reveal characteristics unique to these raisins. Note the square-shaped bodies and faces, the teeth are more defined, and most importantly, they have lined, bushy eyebrows. Each figure is dated 1987, made in China.

Del Monte musical sandwich, 6" long x 5" wide, made of plastic, which supports three moveable 2.5" high PVC raisins. A button in the corner activates the music box to play *I Heard It Through the Grapevine*. Sandwich is marked: The California Raisins™ © 1987 CALRAB, Claymation® designed by Will Vinton, China. Value: complete MIB $70-75; as shown, no box $30-40.

13

Chapter 2

Figural Characters/Hardee's

Hardee's Restaurant First Issue: *Sax Player, Ben Indasun, Tiny Goodbite, Justin X. Grape.* Each PVC is 2" high, copyright 1987 CALRAB, made in China. In addition to being smaller than the original commercial issue, these raisins have different hand positions and facial expressions. Value $3-5 each.
Tip: These four figures were issued commercially as keychains, so if they are found with a small hole in the top of the head, then you have a keychain, missing the chain.

Hardee's Restaurant Second Issue consists of six PVC characters, ranging from 2" to 2.25" high, copyright 1988 CALRAB, manufactured by Applause™, made in China. From left: Waves Weaver, Trumpy TruNote, Rollin' Rollo, Captain Toonz, S. B. Stuntz, and F. F. Strings.

Waves Weaver, $5-7; MIP trading card, $8-10

Trumpy TruNote, $5-7; MIP trading card $8-10.

F. F. Strings, $5-7; MIP trading card, $8-10.

S. B. Stuntz, $5-7; MIP trading card, $8-10.

Rollin' Rollo, $5-7; MIP trading card $8-10.

Hardee's Restaurant Third Issue was a group of four soft raisins, with poseable arms and legs. Each is approximately 5.5" long, copyright 1988 CALRAB, licensed by Applause, made in China. Shown here in original plastic packaging. Value MIP $10-12 each.

Captain Toonz, $5-7; MIP trading card $8-10.

Female soft raisin, item #30087, wearing yellow shoes. **Conga Dancer** soft raisin, item #30085, wearing blue sneakers and yellow cap with a red letter "H." **Microphone** soft raisin, item #30084, with microphone sewn into his left hand. **Sunglasses** soft raisin, item #30086, wearing orange sunglasses and sneakers. Value $6-8 each.

Hardee's Restaurant Fourth Issue consists of Benny (Ben Indasun), Anita Break, Alotta Stile, and Buster, all released in 1991. Each PVC came in a plastic package with trading card. All are copyright CALRAB, licensed by Applause, and made in China. Value MIP $15 each.

Benny, 2.25", is our old friend Ben Indasun, who enjoys bowling at the Fresno Grapevine Bowl. **Anita Break**, 2.25", is Benny's wife. Anita likes to shop and often stops at Hardee's to pick up dinner for her family. **Alotta Stile**, 2.25", is Benny and Anita's daughter, who often stops at Hardee's and likes to listen to her favorite singers, "New Grapes on the Vine." **Buster,** 2", is Benny and Anita's son. With his lightning skateboard, Buster likes to have fun and go with the flow. Value $8-10 each.

Chapter 3

Figural Bendees & Walkers

Bendee's, shown from left are Ben Indasun, Tiny Goodbite, and Justin X. Grape. Each has flexible arms and legs, so poses may vary; 5.5" high; the hard PVC body is flat and round like a pancake. Each figure is marked on the bottom of one shoe: The California Raisins™, © 1987 CALRAB, Applause, China. Value $15-18 each.

Walker's, shown from left are Justin X. Grape, Tiny Goodbite, and Ben Indasun. These plastic windup walkers are 3.5" high; they have moveable arms, so poses may vary; marked: "© 1988 CALRAB, Licensed by Applause Licensing, Nasta Ind. Inc. N.Y., Made in China." Value $12-15 each.

Tiny Goodbite walker in blister pack. Value $25.

Ben Indasun walker in blister pack. Another MIP Ben walker has an original price tag of $2.49 from Zayre. Value $25.

Orange sunglasses walker on left is the 1987 original-design walker at 2.75" high, flip-up sunglasses, made smaller and more fragile than the later walker. Marked: "The California Raisins™, © 1987 CALRAB, Applause®." On the right is the more commonly found 3.5" 1988 walker.

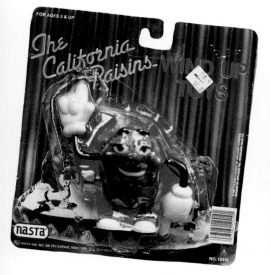

Justin X. Grape walker in blister pack. Note the $2.86 price tag from K-Mart. Value $25.

Orange Sunglasses walker, original design, with sunglasses lifted to give a view of his eyes. Value $45-50.

Green Shoes Female (Ms. Delicious) wind-up walker in blister pack; each shown is marked: "© 1988 CALRAB, Licensed by Applause, Nasta Ind. Inc. N.Y., Made in China." Value $35.

Green Shoes Female (Ms. Delicious) walker with tambourine, 3.75" high. Value $20-25 with; $10-12 without.

Pink Shoes Female (Ms. Sweet) walker with tambourine, 3.75" high. The tambourine is a separate piece that was inserted into a tiny pinhole in right hand and affixed. All too often the tambourine is missing, thus the walker is considered damaged. Value $20-25 with; $10-12 without.

Yellow Shoes Female (Ms. Marvelous) walker with tambourine, 4" high. Value $20-25 with; $10-12 without.

18

Chapter 4
Backpacks, Tote Bags, Purses, & Wallets

Original blue package, backpack is blue with orange trim. Value $35-45.

3 Raisin Men backpack, triangular, blue with orange bottom, original tag. Value $35-45. (*Courtesy of Barry Ullmann*)

Original yellow package, backpack is blue with orange trim. Value $35-45.

Original blue package, backpack is yellow with maroon trim. Value $35-45.

The Rayettes, female backup singers on stage, turquoise backpack with yellow trim. Value $30-35. (*Courtesy of Barry Ullmann*)

3 Raisin Men tote, blue canvas, original wrapper and tag, 11.5" square. Value $20-25. (*Courtesy of Barry Ullmann*)

Conga Dancers backpack, adjustable padded shoulder straps, carrying handle. Value $22-25.

Tiny Goodbite and a Rayette, with zippered outer pouch. Value $30-35.

3 Conga Dancers tote, yellow canvas, 11.5" square. Value $15-18.

Wallet. Value $12-15.

Fanny Packs, two styles shown. Value $15-18 each. (*Courtesy of Barry Ullmann*)

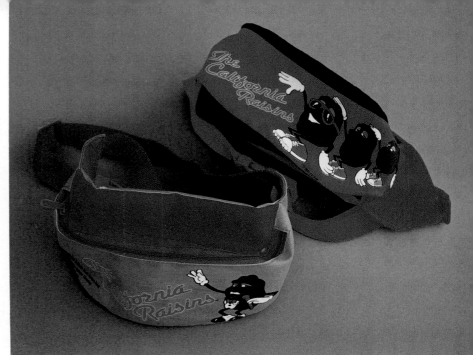

Wallet, child's, mint in blister pack. Value $20-22. (*Courtesy of Barry Ullmann*)

Purse. Value $20-25

3 Conga Dancers tote, pink canvas. Value $15-18.

3 Conga Dancers tote, yellow canvas. Value $15-18.

Chapter 5

Books & Bookmarks

What's Cool! Copyright 1988 CALRAB, Applause, by Checkerboard Press of Macmillan, Inc., 8.25" x 5.25", color paperback cover, black and white inside pages. Value $7-9.

What's Hot! Copyright 1988 CALRAB, same as previous book. Value $7-9.

Raisins In Motion (A flashback). Copyright 1988 CALRAB, Applause, hard cover book by Checkerboard Press of Macmillan, Inc., all color. Value $15.

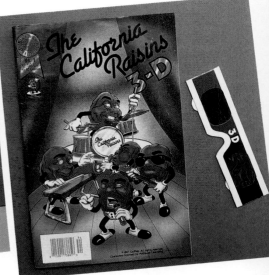

Raisins In Motion (A flashback). Copyright 1988 CALRAB, Applause, Checkerboard Press, 8" x 8.25" paperback cover, all color. Value $10-12.

Birthday Boo Boo. Copyright 1988 CALRAB. Value $10-12.

The California Raisins™ 3-D, copyright 1988 CALRAB, Applause, 8.75" x 10.25" book, Blackthorne Publishing, Inc., paperback, shown here with 3-D glasses, No. 1 in series. Value $8-10.

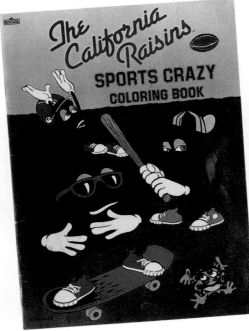

A Haunting We Will Go! Copyright 1988 CALRAB. Value $10-12.

Raisin' The Roof. Copyright 1988 CALRAB. Value $10-12.

Sports Crazy coloring book. Copyright 1988 CALRAB, Applause, by Marvel Books of Canada; 8" x 11" with color cover, black and white pages for coloring. Value $20 uncolored.

The California Raisins™ 3-D Hollywood, No. 2 in series. Value $8-10.

Coloring books, *Go to College, Vacation Fun in the Sun, On Tour*, all same as *Sports Crazy*. Value $20 each uncolored.

23

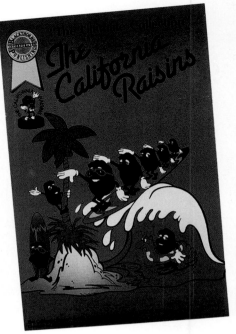

The California Raisins™ *3-D,* No. 3 in series. Value $8-10.

The California Raisins™ *3-D,* No. 5 in series. Value $8-10.

The Ultimate Collection, copyright 1988 CALRAB, Applause, Blackthorne Publishing, Inc., larger comic paperback. Value $15-20.

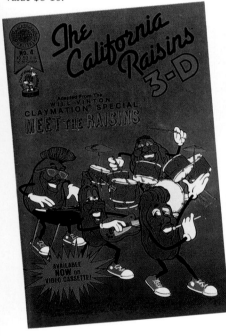

The California Raisins™ *3-D Meet The Raisins*, No. 4 in series. Value $8-10.

Animated Sticker Album, copyright 1988 CALRAB, Applause, by Diamond Publishing, Inc., 8.5" x 11" paperback. Value $5-7.

Bookmark with tassel, © 1987 CALRAB, Applause, 1.5" x 5 7/8", Antioch Publishing Company. Value $4-5.

Bookmarks with tassels, same as previous. Value $4-5 each.

Bookmarks without tassels, same as previous. Value $2 each.

Bookmarks with cutout tab to hook over book page. Value $3 each.

Bookmarks with cutout tab. Value $3 each.

Gift Enclosure Bookmark, © 1988 CALRAB, Applause, 1.5" x 5.5" paper. Value $2.

Gift Enclosure Bookmark. Value $2.

Gift Enclosure Bookmark, shown from back.

Bookmarks with cutout tab. Value $3 each.

Chapter 6

Clocks & Watches

Wristwatch Wall Clock, red plastic, 1988, 36" long, yellow and white face with singer reads "Keeping Time with ... *The California Raisins™.*" Value $50-75. (*Courtesy of Barry Ullmann*)

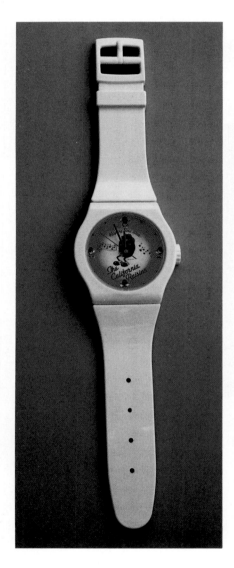

Wristwatch Wall Clock, white plastic, same as previous. Value $50-75. (*Courtesy of Barry Ullmann*)

Wristwatch Wall Clock, red plastic, 1988, 36" long, four raisins on stage with blue curtain. Value $50-75. (*Courtesy of Barry Ullmann*)

Wristwatch Wall Clock, blue vinyl strap, 1988, 54" long, four raisins on stage with blue curtain. Value $75-85. (*Courtesy of Barry Ullmann*)

Closer view of previous blue-strap wristwatch face.

Wristwatch Wall Clock, white plastic, same as previous. Value $50-75. (*Courtesy of Barry Ullmann*)

Fan Club Watch Set, © 1987 CALRAB, Applause, by Nelsonic, watch came with black, white, and dark purple interchangeable watchbands. Fan club membership was $5.95 and three proofs of purchase from any brand, any size, of California raisins; watch set was sent in membership package. Value $20-25.

Watches of different lengths. Compare the singers' shoes, bow ties, dickeys, facial expressions, hands, etc. Value $20-25 each.

Singer lifted to reveal digital watch.

Wristwatch, conga dancers pictured on face, colorful musical notes on band. Value $20-25. (*Courtesy of Barry Ullmann*)

Wristwatch, black leather, raisin surfing a wave on watch face, marked *The California Raisins*™." Value $18-22. (*Courtesy of Stan, Dayle, & Allen Golomski*)

Wristwatch, white, 7.5" watch, © 1988 CALRAB, Applause, Nelsonic, white band with plastic conga dancer raisin watch. Value $40-45 MIP.

Chapter 7
Clothing & Costumes

Perhaps more than any other category, clothing prices are dependent on condition. Listed book prices should not appear on items that have stains, rips, tears, fading, stretching, or any combination of these, as they will have little value to a collector. Halloween costumes may have only been worn once or twice during one holiday season, so these may be found in better condition than everyday clothing. Christmas clothing also had a short season to be worn, so condition is generally good on these items. Raincoats, hats, belts, and the rest, are expected to be found with normal wear. Don't always count on finding items in the mint condition we've been able to picture.

Hardee's jacket and cap, shown from back.

Hardee's Jacket and Baseball Cap, shown from front, in purple satin with knitted cuffs and collar. Value $150-200 Mint. (*Courtesy of Barry Ullmann*)

Girl's Raincoat with conga dancers. Value $20-25. (*Courtesy of Barry Ullmann*)

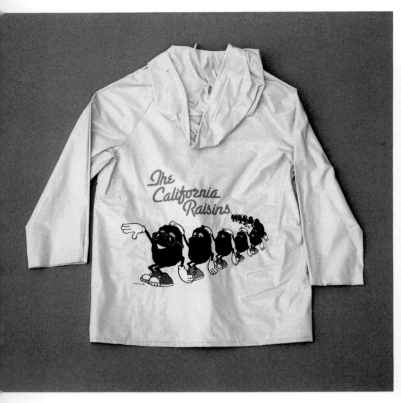

Girl's Raincoat with conga dancers. Value $20-25. (*Courtesy of Barry Ullmann*)

Child's Raincoat with conga dancers. Value $20-25. (*Courtesy of Barry Ullmann*)

Girl's Raincoat with female raisins and boys at beach. Value $35-40. (*Courtesy of Barry Ullmann*)

Neckerchief, 23" x 24", copyright CALRAB. Value $10-12.

Boxer Shorts (All boxer shorts shown are adult sizes.) Value $8-10.

Denim Skirt, girls size 7. Value $12-15.

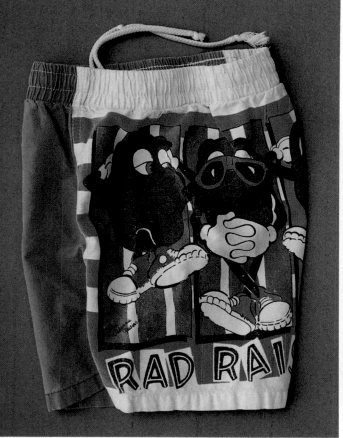

Rad Raisins swimming trunks, child's,
copyright 1987 CALRAB. Value $12-15.

Boxer Shorts. $8-10.

Christmas Boxer Shorts. $10-12.

Christmas Boxer Shorts. $10-12.

Girls' Sneakers and Slippers. Each $10-12 Mint ; $5-7 used.

Tennis Wrist Band, printed figure of *A.C.* Value $3-5.

Baseball Caps with conga line. Also found with Microphone Singer on cap. Value $12-14 each.

Ear Muffs, six sets shown, all with original tags. Value $10-12 pair, mint; $4 used.

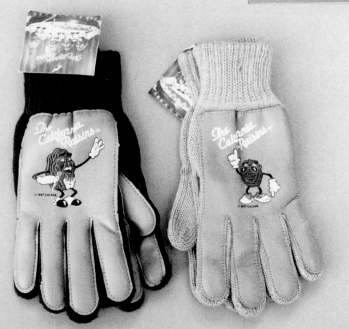

Gloves, two colors shown, each with original tag. Value $15 pair, mint; $4 used.

33

Visor, white plastic, © 1987, with conga line. Value $5-7.

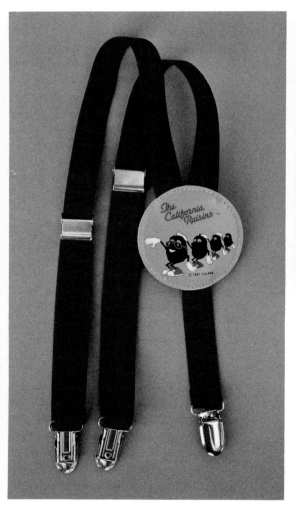

Elastic Belts, mint on holder, three styles shown, made by *Lee*. Value $12-15 each. (*Courtesy of Barry Ullmann*)

Elastic Belt with metal clasp, original foil label. Value $10-12. (*Courtesy of Robert & Lois Duvall*)

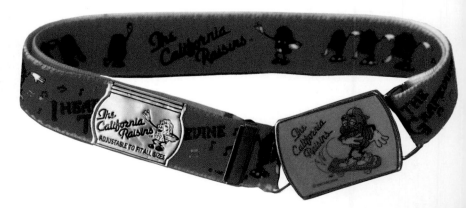

Suspenders © 1987. Value $18-20. (*Courtesy of Barry Ullmann*)

Painter's Caps. Value $12-14 each.

Corduroy Cap, offered to fan club members for $5.50 postage paid. Value $15-18.

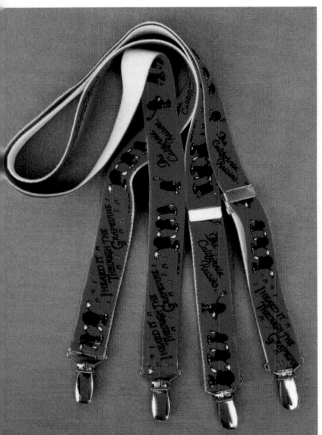

Elastic Belts with plastic raisin figure clasp. Value $10-12 each.

Suspenders ©1987. Value $18-20. (*Courtesy of Barry Ullmann*)

Elastic Belt with metal clasp, original foil label. Also found solid blue elastic belt with same clasp. Value $10-12. (*Courtesy of Robert & Lois Duvall*)

Halloween costume and box, Collegeville, of male raisin, with white gloves. These costumes were open at the base and slid over the head; eye holes can be seen between the blue eyelids on costume, and breathing holes are between the eyes. There is a clear plastic eye mask with elastic band attached to the inside of the costume at the eye openings. To complete the costume, person was required to add their own black pants or tights, long sleeve black jersey, and shoes. Value $25-30 MIB.

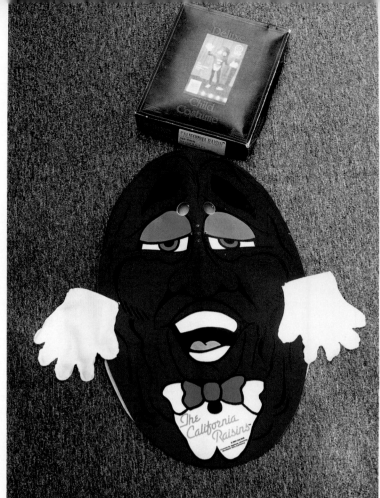

Suspenders, on original hanger, © 1987. Value $22-25. (*Courtesy of Barry Ullmann*)

Halloween costume on original plastic hanger, made by *Cover Ups*. Value $25-30.

36

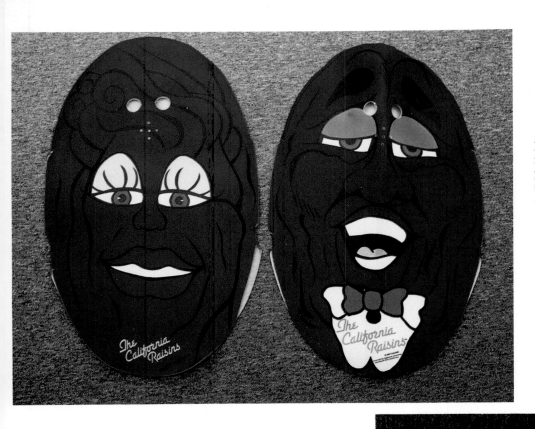

Halloween Costumes of female raisin and male raisin; white gloves not shown; each produced in adult and children's sizes of small, medium, large. Female costumes will average slightly more than male costumes.

Post™ Raisin Bran T-shirt was available free by sending in eight proofs of purchase; or by sending in two proofs of purchase and $4.99. Came in sizes S, M, L, XL. Value $25-30 Mint; $12-15 Used. (*Courtesy of Toni Crittenden & Art Voorhees*)

Child's Sweatshirt. $5-8.

Talent Night adult sweatshirt. $15-18.

Play in the Bowl of Raisins adult sweatshirt. Value $15-18. (*Courtesy of Stan, Dayle, & Allen Golomski*)

I Heard It.... sweatshirt. Value $15-18. (T-shirt found with same design. Value $10-12.)

On Tour/Sold Out sweatshirt. Value $15-18. (T-shirt found with same design. Value $10-12.)

Yule Tied Christmas sweatshirt. $20-22.

Child's Sweatshirt. $5-8.

Jingle Bell Rock Christmas sweatshirt.
$20-22.

39

Do-Woping in the Snow sweatshirt. $20-22.

Be-Boppin hooded cover-up. $15-18.
(*Courtesy of Barry Ullmann*)

A Day at the Beach T-shirt. $10-12. (*Courtesy of Barry Ullmann*)

All-Stars T-shirt. $10-12.

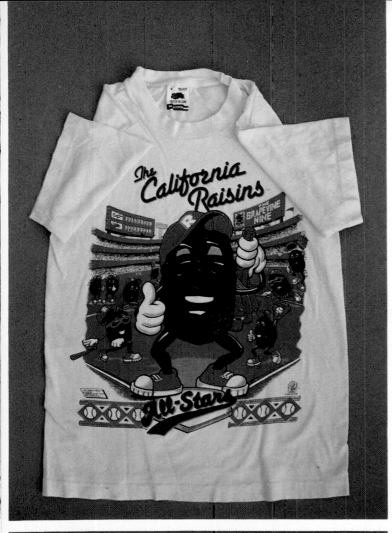

Rapid Transit! T-shirt. $10-12. (*Courtesy of Barry Ullmann*)

Summer '88 T-shirt. $10-12.

Grapevine Spike Time tank top. $12-15.

Greetings! T-shirt. $10-12.

Jersey. $8-10.

Crop-top jersey. $8-10. *(Courtesy of Stan, Dayle, & Allen Golomski)*

Rock 'n Raisins jersey. $10-12. *(Courtesy of Stan, Dayle, & Allen Golomski)*

Child's T-shirt. $5-7.

Pajama Top. $3-5.

Nightshirt. $12-14.

Nightgown. $12-14.

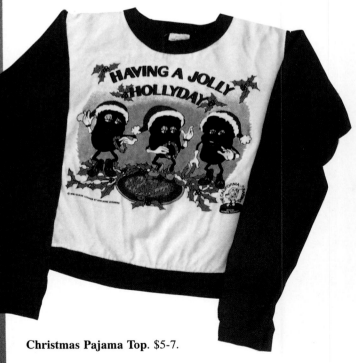

Christmas Pajama Top. $5-7.

Chapter 8

Cups, Glasses, & Mugs

The California Raisins™ glassware, all with a conga line of singing raisins, shown from left: **Vase** (or cocktail shaker), 7", $20-25; **Tumbler** with ribbed base, 5.75", $12; **Tumbler**, 5.25", $12; **Juice** glass, 4", $12; and Beverage **Decanter**, 8", or 8.75" with lid on, $25-30.

Label on bottom of ribbed tumbler.

Candy Container with plastic lid, glass is 8" high, 4" diameter. Value $25-30.

Ribbed cups, plastic, with Tiny Goodbite, Ben Indasun, and Justin X. Grape, 22 ounce, © 1987. Value $10-12 each.

Hardee's Grapevine Tour '88 plastic cups, 4.25", 12 ounce, set of four. Shown at top left is one cup showing the front, others are turned to the different back scenes. Value $3-4 each.

Mini Max mug on left, plastic, 3.5" high as shown. Blue base could be affixed to surface of a vehicle for stability, with white removable mug; blue lid is non-splash with small sipping hole; same raisin scene is printed on both sides. Value $8-10 complete. **Super Max** mug on right is plastic, 20 ounce, 5", white with blue non-splash lid; same raisin scene is printed on both sides. Value $8-10.

Super Max mug, plastic 20 ounce, 5" high, white with orange non-splash lid; microphone singer on one side, with stage singers and conga line scene (see previous Max mugs) on opposite side. Value $8-10.

Sports Mug with plastic straw, shown on left, 32 ounce, 6.5", handle, light gray with purple logo, made by Enduro. Value $18-20 complete. **Sports Bottle** with plastic straw, 8.5" high, conga dancers. Value $12-15 complete.

Super Mug, shown 7.5", 6.5", and 5.25" high, plastic with ribbed base. Values: Left $12; center $10; right $7.

Different view of two previous sports containers.

California Gold cup, copyright 1987
CALRAB. Value $8-10.

Back view of box.

Christmas 1988 Collector's Mugs, mint in
box, limited edition mugs offered to fan
club members for $15.85 postage paid. Box
is 15.5" high with 3" top panel. Value: $90-
100 Set of four mugs MIB; $20 each mug,
no box.

Views of the four Christmas 1988 mugs, each mug in the same position but turned to show the beautiful scenes.

Raisins on Skis, same mug shown both sides, 3.75" high, copyright 1988 CALRAB. Value $15.

Birthday mug, 3.75", copyright 1987 CALRAB, shown with original box. Value: $18-20 mug MIB; $15 for mug only.

Inside view of birthday mug.

Winter Scene, 3.75", copyright 1988 CALRAB, raisin conga line dancing through snow. Value $15.

Reverse side of previous Winter Scene mug.

Mugs, three variations, 3.75" high, copyright 1987 CALRAB, themes from *"I Love You,"* to *"Happy Birthday,"* to *"You make me feel like dancing."* Value $15 each

Valentine mugs, set of four, 3.75",
copyright 1988 CALRAB. Value $20-25
each. (*Courtesy of Barry Ullmann*)

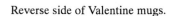

Reverse side of Valentine mugs.

Christmas mug on left, **Winter Scene** mug
with ice skaters on right, 3.75", copyright
1988 CALRAB. Value $15 each.

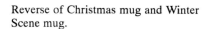

Reverse of Christmas mug and Winter
Scene mug.

Chapter 9

Displays & Promotional Items

Hardee's display stage made of heavy cardboard, 18" x 22", purple stage in front is 4" high. The raisins featured are Hardee's first series, and they dance on stage powered by two 'D' batteries. The line of three raisins in back glides from side to side, while the singer in front swivels back and forth in a half circle. Value $250 +.

Thrifty Maid grocery store cardboard display, 62" x 20.5", for displaying raisin packages, copyright 1987 CALRAB. Value $100-110.

Closer view of Hardee's stage, showing the plastic battery compartment, which tucks under the stage. Note the domed cutout behind the raisin figures with a view of the blue stage curtain. The curtain is a separate sheet of heavy paper, held to the back of the display. If this curtain were missing the back would be open.

Display Stage used in gift/card/retail shops to display commercial PVC raisins. Vacuform stage is 13.25" x 17.75", the white stage area is 9.5" long, and the peach tray area in front is 2.5" high. Value $100-125. (*Courtesy of Toni Crittenden & Art Voorhees*)

From the Heart Valentine display box, shown closed, came with 36 PVC figures, 18 male and 18 female. Cardboard box with cutout cover for store display, 9.5" x 7.25" x 4.25", copyright 1988 CALRAB. Value box only $40-50.

Valentine display box open; PVC items for display only.

Christmas Figurines display box, shown closed, cutout cover for store display, 9.5" x 7.25" x 4.25", copyright 1988 CALRAB. Value $40-50.

Raisin' Some Fun in the Sun! display box, shown closed, cutout cover for store display, 12.25" x 8" x 4.5", copyright 1988 CALRAB. Value $40-50.

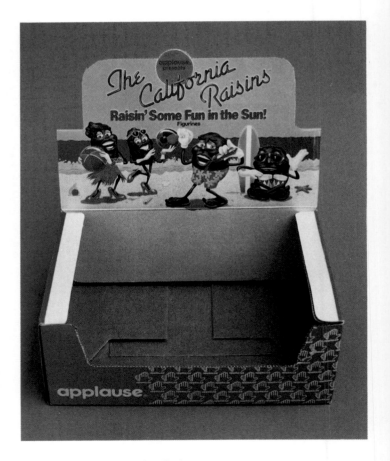

Christmas Figurines box open for display.

Fun in the Sun! box open for display.

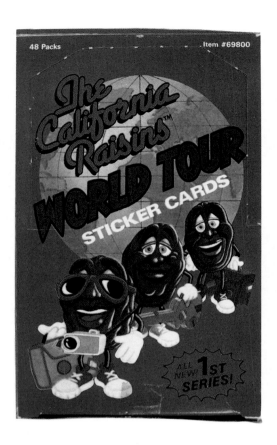

The California Raisin Show display box, shown closed, cutout cover for store display, 8" x 7" x 3.5", © CALRAB. Value $40-50.

World Tour Sticker Cards display box containing 48 packs of sticker cards, shown closed, cutout cover for store display, 7" x 4.5" x 2", copyright 1988 CALRAB. Value box only $15-20; sealed packs $5 each.

Raisin Show box open for display.

World Tour box open, displaying mint packages of sticker cards.

World Tour packs of five cards come in blue wrappers, shown front and back. Sticker cards represent twenty-four countries all over the world, with the United States, Spain, and Switzerland cards pictured. Value: sealed pack of five cards $5; single cards $1 each.

Raisin Man 3-D Vacuform display, 30" x 30", for wall or post display. Value $150-175. (*Courtesy of Barry Ullmann*)

Ear Muff display box, 19.5" x 14.5" x 3", copyright 1987 CALRAB. Value $35-40.

Auto Sun Shield display, 24" x 18", copyright 1987 CALRAB. Value $30-35.

Gift Center mobile display, 22" x 18", copyright 1987 CALRAB. Value $30-35.

Cream of Wheat store ad display board, pictured with a 28-ounce box of Cream of Wheat which has a California Raisin on front. Board Value $20-25.

Sax Player mobile display for party supplies, same scene both sides, 14" x 19", copyright 1988 CALRAB. Value $12-15.

Coupon which was on the Cream of Wheat store display. Value $2-4.

Chapter 10

Keychains

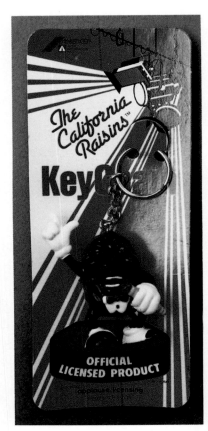

Tiny Goodbite keychain, mint on card, card is 6.75" long, the entire keychain is 4.25" long, copyright 1987 CALRAB, elastic band holds the PVC figure to card. Value $12 MOC; $5-7 loose.

Justin X. Grape keychain. Value $12 MOC; $5-7 loose.
Tip: These PVC items are the same as the first Hardee's release, with a keychain added. Check closely when buying these, and if you find a PVC with a small hole in the top of the head, you're buying a former keychain without the chain. Value as Hardee's releases (no hole in head) is $3-5 each.

Sax Player keychain. Value $12 MOC; $5-7 loose.

Ben Indasun keychain. Value $12 MOC; $5-7 loose.

Set of Four PVC keychains, mint in blister pack, card is 6.75" long, same keychain as on previous cards, copyright 1987 CALRAB. Value MOC $20 each.

Justin X. Grape Graduate keychain, entire length approx. 4.75", PVC figure 2 3/8" long, copyright 1988 CALRAB. Value $50-75. (*Courtesy of Barry Ullmann*)

Tiny Goodbite Graduate keychain, entire length approx. 4.75", PVC figure 2-3/8" long, copyright 1988 CALRAB. Value $50-75. (*Courtesy of Barry Ullmann*)

Hip Band keychains, set of four, shown from left: Hip Guitarist, Sax Player with Black Beret, Girl with Microphone, Boy with Microphone. Each keychain is 4.25" long, PVC figure 2" long, slight design changes from the figurines. Value $20-25 each.

Ben Indasun Graduate keychain, entire length approx. 4.75", PVC figure 2-3/8" long, copyright 1988 CALRAB. Value $50-75. (*Courtesy of Toni Crittenden & Art Voorhees*)

Sax Player Graduate keychain, entire length approx. 4.75", PVC figure 2 3/8" long, copyright 1988 CALRAB. Value $50-75. (*Courtesy of Toni Crittenden & Art Voorhees*)

Surfboard metal keychain, approx. 3" total length, copyright 1988 CALRAB, Applause, made in Taiwan. Value $8-10.

Microphone Singer metal keychain. Value $8-10.

Conga Dancer metal keychain. Value $8-10.

"Ooohoo" Singers metal keychain. Value $8-10.

Straw Hat & Cane metal keychain. Value $8-10.

Conga Dancer metal keychain. Value $8-10.

61

Four Surfers in Wave plastic keychain with paper insert, scene is the same on both sides, total 3.5" long, copyright 1988 CALRAB, Applause, by *Button-Up* of Troy, Michigan. Same information applies to each of the following variations of this keychain. All shown valued at $4-6 each.

Surfer on Beach and **Tub Scene**.

Conga Line and **Surfer in Wave**.

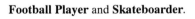

Football Player and **Skateboarder**.

Hawaiian Conga Line and **Sunbathers on Towels**.

Microphone and **Sax Player**.

"Ooohoo" Singers and **Microphone & Singers**.

Golfer and **Conga Dancer**.

Chapter 11

Linens: Bedding, Kitchen, & Towels

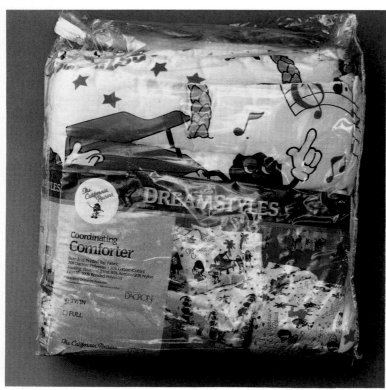

Comforter, available in full and twin sizes, copyright 1988 CALRAB, Applause. Value MIP $50-60; used $20-22. (*Courtesy of Barry Ullmann*)
Note: As with clothing, used linens should have minimal everyday wear and tear. Items in poor condition will command far lower than listed prices.

Bedspread available in full and twin sizes, copyright 1988 CALRAB, Applause. Value MIP $50-60; used $20-22.

Sleeping Bag, child's, with light gray background, copyright 1987 CALRAB, Applause. Value MIP $40-45; used $18-20.

Bed Pillow shown mint in package, copyright 1988 CALRAB, Applause. Value MIP $35-40; used $8-10. (*Courtesy of Barry Ullmann*)

Sheets available in full and twin sizes, copyright 1988 CALRAB, Applause. Value of complete set, MIP $25-30; used $10-15 set.

Beach Throw Blanket, shown mint in tote bag, 45" x 72", copyright 1988 CALRAB, Applause. Value MIP $40-45; used $18-20. (*Courtesy of Barry Ullmann*)

Waterbed Sheet Set available in king, queen, super single, copyright CALRAB, Applause. Value MIP $40-45; used $15-25. (*Courtesy Stan, Dayle, & Allen Golomski*)

Reverse side of waterbed sheet set.

Draperies, pinch-pleated, 48" x 63", copyright 1988 CALRAB, Applause. Value $20-25 pair.

Surf's Up! beach towel, 30" x 60", copyright 1989 CALRAB, offered to fan club members for $10 postage paid. Value $35-45 Mint. (*Courtesy of Toni Crittenden & Art Voorhees*)

Sunbathers on Towels beach towel, 27" x 52", copyright 1988 CALRAB. Value $15-20.

Diner Waitresses potholder, 7.75" x 7.75", copyright 1988 CALRAB. Value $4-5.

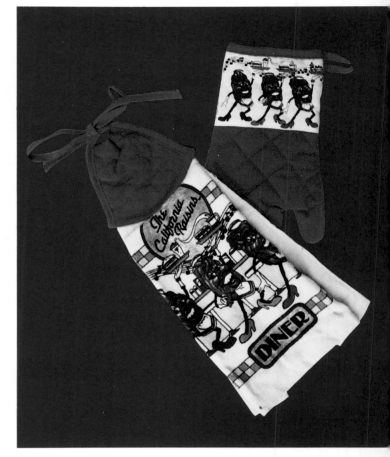

Diner Waitresses hand towel with tie, and oven mitt, copyright 1988 CALRAB. Value $6-8 each.

Diner Waitresses dish towel, 16" x 24", copyright 1988 CALRAB. Value $6-8.

Raisin Chefs potholder 7.75" x 7.75", dish towel 16" x 24", dish cloth, copyright CALRAB. Value, left $4-5, center $6-8, right $6-8. (*Courtesy of Barry Ullmann*)

Chapter 12

Magnets

PVC Magnets, set of four, copyright 1988 CALRAB, made in China. Value $20-25 each. (*Courtesy of Barry Ullmann*)

PVC magnets shown from back.

Though these PVC magnets look so 3-D, when turned sideways you can see how incredibly flat-backed they are. (*Courtesy of Allen Golomski*)

Conga Dancer, metal, 1.5" diameter. Value $3-5.

Note: This magnet, and the metal ones shown following in this chapter, were also produced as pins.

Football Player. $3-5.

Sax Player. $3-5.

Hiker. $3-5.

Tennis Player. $3-5.

Conga Line. $3-5.

Conga Dancer. $3-5.

Conga Dancer. $3-5.

Microphone Singer. $3-5.

Back of magnet.

Jogger. $3-5.

Chapter 13

Miscellaneous

Animation Cel from Will Vinton Studios, produced for *The California Raisins™ Show*. Cel is 9" x 10.5" with background, includes Certificate of Authenticity (framed by owner). Value $150-175. (*Courtesy of Barry Ullmann*)

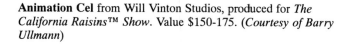

Animation Cel from Will Vinton Studios, produced for *The California Raisins™ Show*. Value $150-175. (*Courtesy of Barry Ullmann*)

Animation Cel from Will Vinton Studios, produced for *The California Raisins™ Show*. Value $150-175. (*Courtesy of Barry Ullmann*)

Certificate of authenticity for animation cels.

Sunglasses, child's, two raisin figures appear on either red, orange, yellow, or green sunglasses. Value $12-15 MIP; $6-8 loose.

Umbrella, child's, 22" long, opened span of 28". Value $20-25.

Sun-Maid Raisin Bank is plastic with a paper label affixed to the raisin box, 6.75" high. In a 1988 Halloween Sweepstakes by Sun-Maid, 10,000 of these banks were given away as second prizes. Also, in 1989, those sending in five Quality Seals from five bags of Sun-Maid Mini-Snacks received this bank for free, along with a coupon for a free bag of Mini-Snacks. Value $18-22. (*Information Courtesy of Don & April Barcus*)

Amazin' Little Raisin Box musical food and raisin container, 7.25" high, musical chip in lid, polyethylene, made in Taiwan. Offered by mail for $4.49 postage paid and one UPC code cut from any seven-ounce or larger package of California raisins. Value $20-25. (*Information Courtesy of Don & April Barcus*)

Reverse side of food container.

Back of bank; look for examples in good condition, paper label not torn or wrinkled, plug in the bottom should be intact.

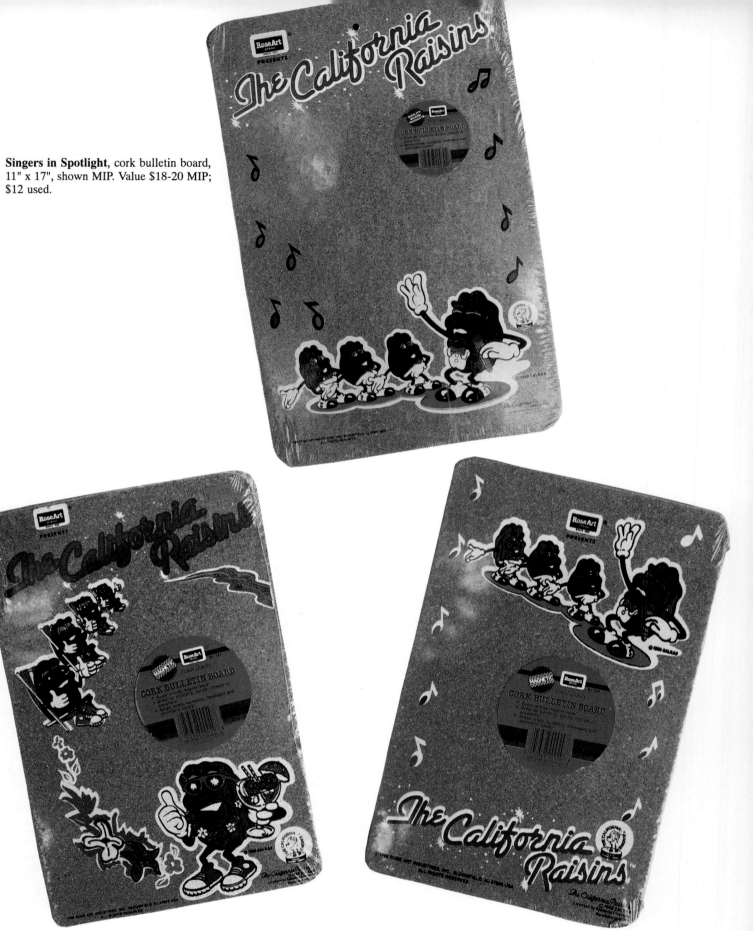

Singers in Spotlight, cork bulletin board, 11" x 17", shown MIP. Value $18-20 MIP; $12 used.

Hawaiian Beach Scene cork bulletin board, 11" x 7.5", shown MIP. Value $12-15 MIP; $8 used.

Singers in Spotlight, cork bulletin board, 11" x 7.5", shown MIP. Value $12-15 MIP; $8 used.

Wipe-Off Memo Board featuring the Rayettes, plastic, 11" x 7-3/8", shown MIP. Value $10-12 MIP; $6 used. (*Courtesy of Robert & Lois Duvall*)
Note: The female raisins shown here are, left to right: *Sweet* in pink shoes, *Delicious* in green shoes, *Marvelous* in yellow shoes. In Chapter 1 we pictured the Fan Club issue of the *Ms. Marvelous* figurine with green shoes, and having the side-swept hairdo. Here, *Marvelous* has the same sideswept hairdo, but has now changed to yellow shoes. No *raisinable* explanation has surfaced, but information is correct in both instances.

Wipe-off Memo Board on left has calendar, plastic, 11" x 7-3/8", shown MIP. Value $12-15 MIP; $7 used. **Wipe-off Memo Board** on right is 7.75" x 5", plastic, shown MIP. Value $10-12 MIP; $5 used.

Wastebasket, 10.25" high x 8" diameter, plastic. Value $40-50. (*Courtesy of Barry Ullmann*)

Scrapbook, 15" x 12.25". Value $18-20.

Address Book, 4" x 2.75". Value $18-20.

Pullchain, Conga Dancer. Value $20-25 MIP; $15 loose.

CrossStitch Kit, Justin X. Grape, 5" x 5". Value $18-20 MIP.

Pullchain for ceiling fan or light, Sax Player. Value $20-25 MIP; $15 loose.

Puffy Stick-Ons. Value $10-12 MIP.

CrossStitch Kit, Tiny Goodbite, 5" x 5".
Value $18-20 MIP.

Hot-Spots, iron-on designs, pack 3 1/8" x
13.25". Value $8-10 MIP.

Puffy Stick-Ons. Value $10-12 MIP.

CrossStitch Kit, Ben Indasun, 5" x 5". Value $18-20 MIP.

The Seat-Shirt with raisins playing volleyball. Value $25-35. (*Courtesy of Barry Ullmann*)

The Seat-Shirt with raisins in jeep. Value $25-35. (*Courtesy of Toni Crittenden & Art Voorhees*)

Back of the Seat-Shirt box.

Auto Sun Shields, three variations shown, MIP. Value $12-15 each.

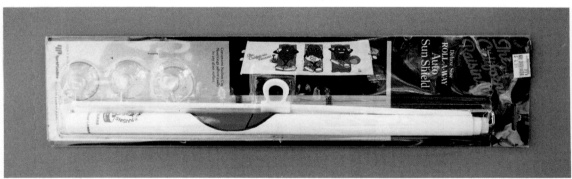

Roll-Away Sun Shield, sunbathers on towels, 51" x 17", shown MIP. Value $35-40.

Sun Shade for rear or side windows, 14.5" x 20", shown MIP. Value $22-25.

Windsical raisin windsock, 44" x 27", shown MIP. Value $15-18 MIP; $8 used.

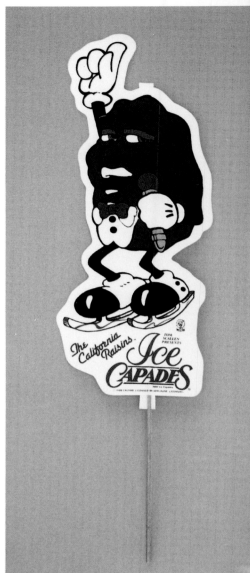

Ice Capades felt pennant on wooden stick, 21.5" long. Value $20.

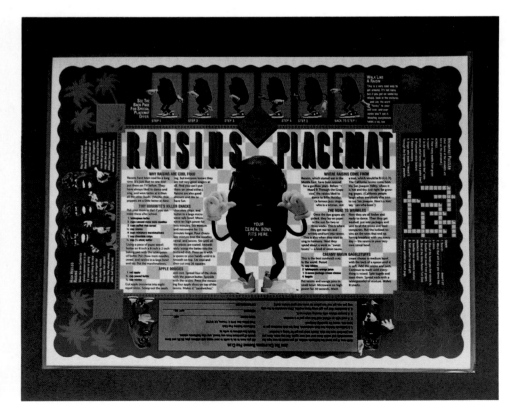

Placemat, 14.75" x 11", laminated; steps on how to walk like a raisin; application to join the fan club; recipes and information. Value $12-15.

Doorknob Hangers, each one shown is 9.25" x 4". Value $5-7 each.

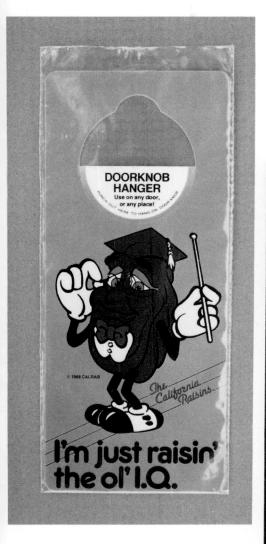

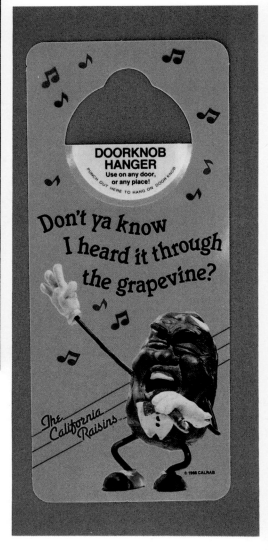

DON'T STAND IN LINE —

Come on in

The California Raisins

© 1987 CALRAB

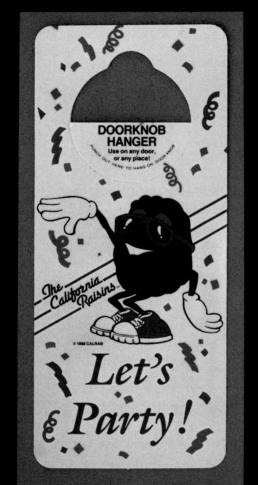

DOORKNOB HANGER
Use on any door,
or any place!

PUNCH OUT HERE TO HANG ON DOOR KNOB

The California Raisins

© 1988 CALRAB

Let's Party!

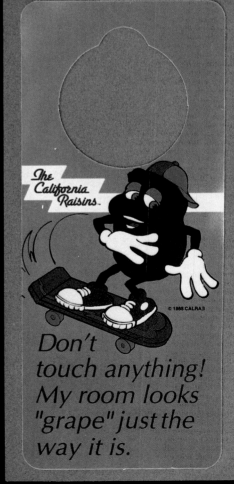

The California Raisins

© 1988 CALRAB

Don't touch anything! My room looks "grape" just the way it is.

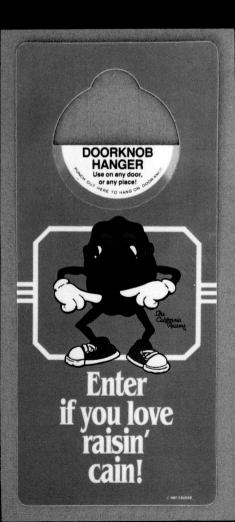

Raisin AM-FM Radio shown in original box, by Nasta, box is 11" x 7.5" and radio is approx. 8" high, uses one 9V battery. Value $125-150.

Raisin radio with microphone, switch for AM-FM on left side, poseable arms and legs, hard plastic body and shoes. Value, no box $75-100.

Back of box.

Back of AM-FM radio.

Raisin AM Radio, microphone missing, poseable arms and legs, hard plastic body and shoes. Value with microphone $110-125.

Meet the Raisins video, copyright 1988 CALRAB, Maier. Value $25-30.

Raisins: Sold Out! video, copyright 1990. Value $25-30.

Back of AM radio.

Meet the Raisins video, copyright 1988 CALRAB, Atlantic, same movie, different box. Value $25-30.

Hip to be Fit video, copyright 1993. Value $15.

Take-a-Long **Cassette player**, in lavender, turquoise, and yellow. Copyright 1988 CALRAB. Value $125-150. (*Courtesy of Barry Ullmann*)

Sing The Hit Songs record album, copyright 1987. Value $12-15.

Sweet, Delicious, & Marvelous record album, copyright 1988. Value $12-15.

Sing The Hit Songs record album which includes a poster and lyrics, copyright 1987. Value $15-20.

Back of album.

Meet the Raisins! record album. Value $12-15.

What Does it Take 45 RPM record in limited edition purple vinyl. Value $25-30. (*Courtesy of Barry Ullmann*)

What Does it Take demonstration 45 RPM record in plain sleeve, copyright 1988. Value $18-20.

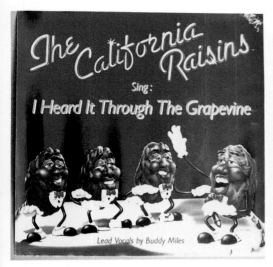

Record 45 RPM, copyright 1987. Value $8-10.

Record 45 RPM, copyright 1987. Value $8-10.

Cassette with original packaging. Value $12-14. (*Courtesy of Toni Crittenden & Art Voorhees*)

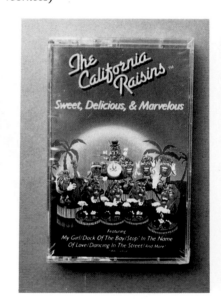

Back of 45 RPM.

Record 45 RPM, copyright 1987. Value $8-10.

Cassette. $10-12.

Record 45 RPM, copyright 1988. Value $8-10.

Cassettes. $10-12 each.

CD shown MIP. Value $30-35 MIP; $15-20 loose.

CD shown MIP. Value $30-35 MIP; $15-20 loose.

Christmas CD copyright 1988. Value $30-35 loose.

Chapter 15

Paper Miscellaneous

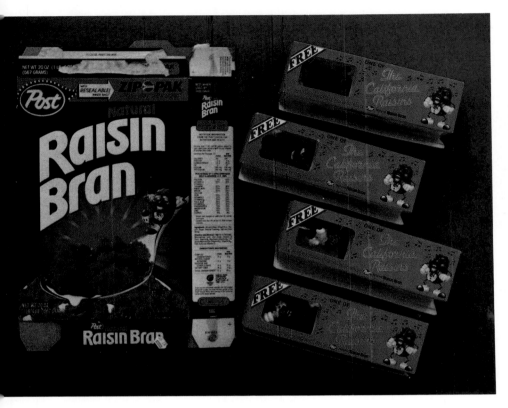

Post Raisin Bran original 20-ounce cereal box, shown from front, with the four boxed PVC raisin figures. One boxed PVC was attached to each box of cereal, free with the purchase of the cereal. Boxes were 7.5" x 2.25" and PVC figures were the four originals: Tiny Goodbite, Ben Indasun, Justin X. Grape, Sax Player. Value of all items shown: $90-100. (*Courtesy of Toni Crittenden & Art Voorhees*)

Post Raisin Bran original cereal box, shown from back, with four mint in package PVC figurines. The offer on the back of the cereal box encouraged consumers to send away for all four PVC's for only $4.95 plus two proofs of purchase, along with the mail-in form on side panel. Value of all items shown: $40-50. (*Courtesy of Toni Crittenden & Art Voorhees*)

Post Raisin Bran cereal box, Canadian issue written in English and French. Value $15-18.

Hardee's Meal Boxes, three variations shown. Value $10-12 each box mint. (*Courtesy of Barry Ullmann*)

Back of Canadian cereal box.

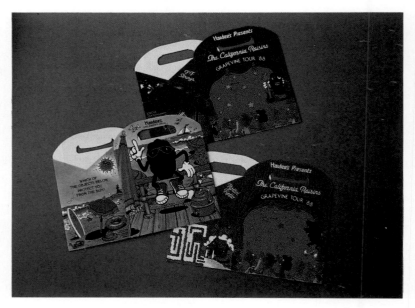

Hardee's boxes, reverse side.

Hardee's placemat for the second series, unused. Value $5-7 mint.

Hardee's ad for the first series. Value $15-20.

Just CrossStitch leaflet #1. Value $10-12.

Leaflet #2. $10-12.

Hardee's paper bag for the fourth series, unused. Value $3-5 mint.

Example shown of Just CrossStitch patterns printed in leaflets.

Leaflet #3. $10-12.

91

Leaflet #4. $10-12.

Postcards, mint condition, never mailed. Value
$2-3 each.

Leaflet #5. $10-12.

Postcard, mint condition, never mailed. Value
$2-3.

Postcards, mint condition, never mailed. Value
$2-3 each.

Postcard of Will Vinton surrounded by
Claymation characters, mint, never mailed.
Value $2-3.

Postcard, mint condition, never mailed. Value
$2-3.

Leaflet #6. $10-12.

Air Freshener mint in package. Value $4-5 MIP; $1-2 loose.

Air Fresheners mint in package. Value for each: $4-5 MIP; $1-2 loose.

Flicker Card of Red, 2.75" x 2", short biography on back. Value $4-6. (Series had four flicker cards; A.C. is missing.)

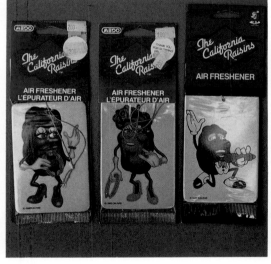

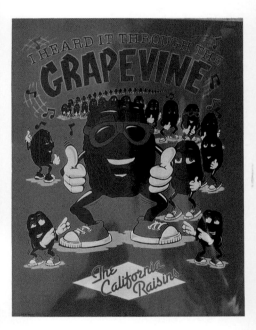

Air Freshener mint in package. Value $4-5 MIP; $1-2 loose.

Air Fresheners mint in package. Value for each: $7-9 MIP for females, $3-4 loose; $4-5 MIP for male, $1-2 loose.

Poster of Conga Dancers, "I Heard It Through the Grapevine", 22" x 28". Value $12-15.

Flicker Cards of Beebop and Stretch, each is 2.75" x 2", copyright 1988 CALRAB, with short biography of each character on back. Value $4-6 each.

Poster of nine-member band. Value $12-15.

Poster of four raisins on stage. Poster came unframed. Value $12-15 unframed.

Reading Club package, complete. Value $25-30.

Poster of Band and Rayettes, 15.5" x 21", unframed. Value $12-15.

Poster, "Books. Check 'em out!", promoting the national reading club for school children, 36" long. Value $12-15.

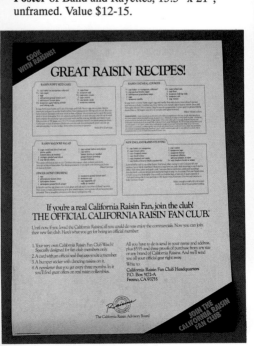

Poster of raisin man happy about breakfast foods with raisins. Value $10-12.

Reverse side of previous poster with four raisin recipes and information for joining the fan club.

Party Supplies: Gift Bag with handle $6-8; Loot Bags in package $4-7; Thank You cards in package $4-7; Party Invitations in package $4-7. All priced mint as shown.

Party Supplies: Hat $1-2; small Balloon $3-4; Cups in package $5-8; Bowls $5-8; large 9" Plates in package $5-8; small Plates in package $5-8. All priced mint as shown.

Tablecloth, paper, yellow, both sides shown mint in package. Value $8-10 MIP.

Napkins in 6.5" and 5" packages. Value $4-7 each MIP.

Wrapping Paper, three colors shown. $8-10 each MIP.

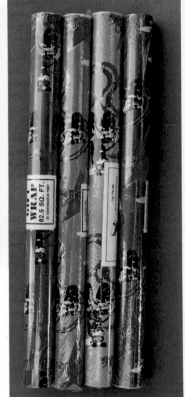

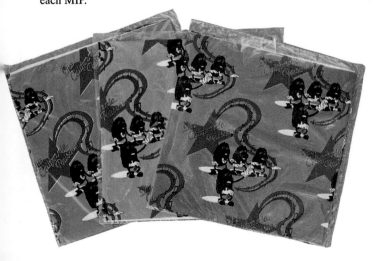

Wrapping Paper, three colors shown. $4-7 each MIP.

Mylar Balloon, large approx. 12". Value $4-5.

Valentine Cards, complete in box. Value $4-6 each MIB.

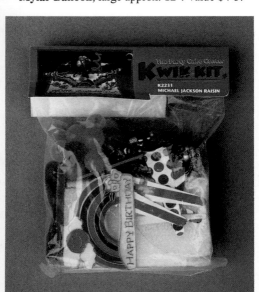

Cake Decorating Kit, Michael Jackson. $30-35 MIP.

Greeting Card. A birthday card is shown, but look for other themes such as graduation cards. Value $2-3 each.

CALRAB Pop'n Jot small Post-It brand notes, cover shown. Value $12-15.

Cake Decorating Kit, Musicians Drummer and Red. $50-55 MIP.

1989 Calendar shown MIP. Value $15-18.

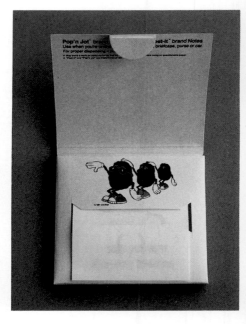

Open Pop'n Jot note papers; you'll see light raisin figure on paper.

Chapter 16

Pins & Buttons

Hardee's plastic pin, 3.75" long. Value $35-40.

Hardee's plastic pin, 3.25" long. Value $35-40.

PVC pin on card, pin is 1.75" long, card is 3.75" long. Value $20-22 MOC; $8-10 loose.

Hardee's name badge. Value $35-40.

PVC pin on card, pin is 1.75" long, card is 3.75" long. Value $20-22 MOC; $8-10 loose.

PVC pin on card, pin is 1.75" long, card is 3.75" long. Value $20-22 MOC; $8-10 loose.

Skateboarder.

Surfer.

Conga Dancer metal cloisonné pin on plastic card, each pin shown is 1.25" to 1.5" long. Value $8-10 each MOC.

Conga Dancer.

Microphone.

Michael Jackson raisin button, 3.5". Value $10-12.

Straw Hat and Cane.

Political Raisins buttons, 1.75". Value $3 each.

Button-Up buttons for remainder of chapter are shown either MOC or loose, all are 1.5" diameter. Value $4 each MOC; $2 loose.

101

Chapter 17

Plush Toys & Stuffed Figures

Thirty-inch male raisin, blue tie, made by Acme, copyright CALRAB. Value $25-30.

Thirty-inch male and thirty-inch female raisins, Acme, copyright CALRAB. Value $25-30 male; $35-40 female.

Twenty-four-inch male and eighteen-inch male raisins, Acme, copyright CALRAB. Value $15-20 each.

Left: 15" male, Acme, copyright CALRAB. Value $12-15. **Center**: 9" male, Acme, copyright CALRAB. Value $8-10. **Right**: 14" musical male, microphone missing in left hand. Value $15.

Left: 20" female, Acme, copyright CALRAB. Value $22-25. **Center:** 15" female, Acme, copyright CALRAB. Value $18-22. **Right**: 10" female, Acme, copyright CALRAB. Value $12-15.

103

Raisin People Pillows, 26" long, satin material, offered to fan club members for $10.50 postage paid. Value $25-30 each.

Fourteen-inch musical male, hard body with stiff arms and legs, vinyl microphone with musical chip. Value $45-50.

Sandals male, 7.5" plush poseable. $10-12.

Grass Skirt girl, 7.5". $10-12.

Microphone, eyes open, 7.5". $8-10.

Blue Guitar, 7.5". $10-12.

Umbrella girl, 7.5". $10-12.

Sax Player, 7.5". $8-10.

Microphone, eyes closed, 7.5". $8-10.

Sunglasses, 7.5". $8-10.

Blue Sneakers, 7.5". $8-10.

Stick-Up in 6.75" x 4.75" package, raisin 6" long, suction cups only on hands. $10-12 MIP.

Stick-Up. $10-12 MIP.

Sunglasses male with suction cups on hands and feet. $6-8.

Stick-Up. $10-12 MIP.

Chapter 18

Raisin Food Products

Sun-Maid raisins in 24-ounce can with Conga Dancer pictured. *Note:* Most of the items shown in this chapter are still available in grocery stores. Look for them before they're gone. No listed prices will be given at this time; they are pictured for information purposes only. You never know where you'll find a *raisinable* piece for your collection!

108

Chapter 19

School Supplies

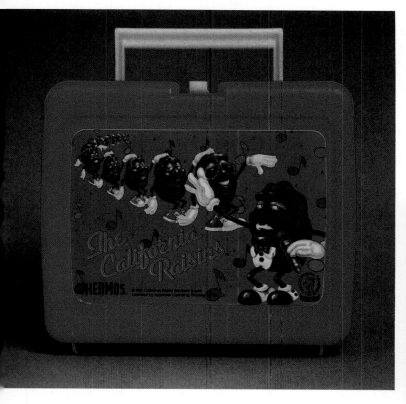

Lunch Box by Thermos, came with raisin thermos. Value $30-40 for mint set.

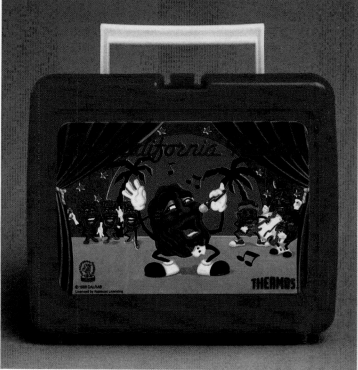

Lunch Box by Thermos, came with raisin thermos. Value $30-40 for mint set.

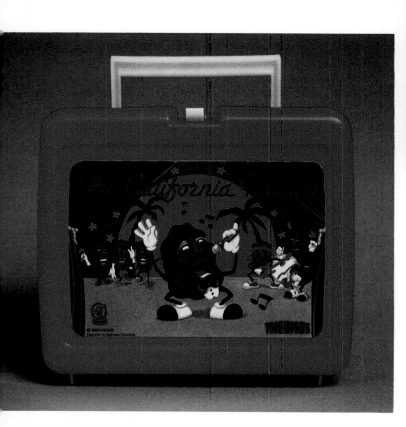

Lunch Box by Thermos, came with raisin thermos. Value $30-40 for mint set.

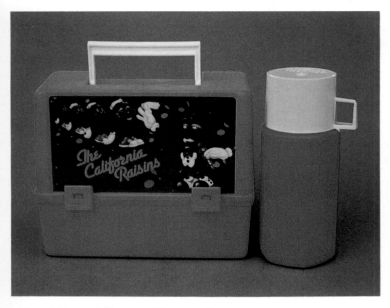

Lunch Box and Hexagon Thermos, Canadian, by Thermos. Value $50 for mint set.

This and the following are examples of thermoses found in the raisins lunch boxes. Some have figural decals, others do not; some have plain screw-on caps under the lid, while others have a cap with spout. Value $8-10 each.

Portfolio shown MIP, showing the two variations produced. Value $25-35.

Portfolio in soft-sided nylon with zipper, 9" x 12", copyright 1987 CALRAB. Value $15-18.

Portfolio open.

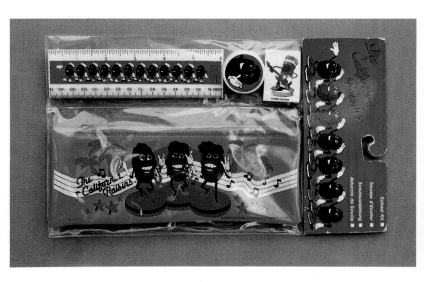

School Kit with pencil pouch, eraser, ruler, pencil sharpener. Value $25-30 MIP.

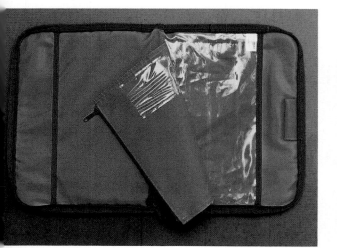

Portfolio open inside.

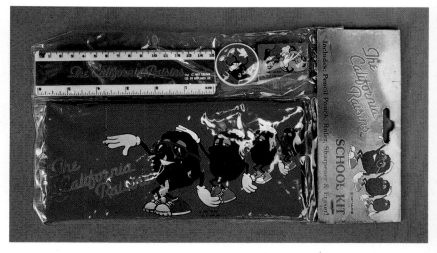

School Kit. Value $20-22 MIP.

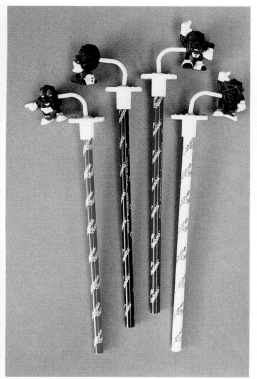

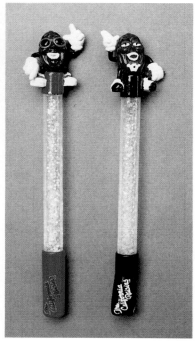

Pencils, set of four with 1.25" to 1.5" PVC figures that twirl on top, copyright 1988 CALRAB. Value $40 unused set.

Glitter Pens with raisin figures, 8" long, copyright 1987 CALRAB. Value $10 each.

Pens, set of three with raisin figures, copyright 1988 CALRAB. Value $30 set.

Erasers, 2.5" high. Value $8-10 MIP.

Folders for 8.5" x 11" papers. Value $3-4 each.

Notebook, 8.5" x 11". Value $8-12.

Picture & Autograph Album, *School Friends*, 6" x 6". Value $25-30.

Chapter 20

Toys, Games, & Puzzles

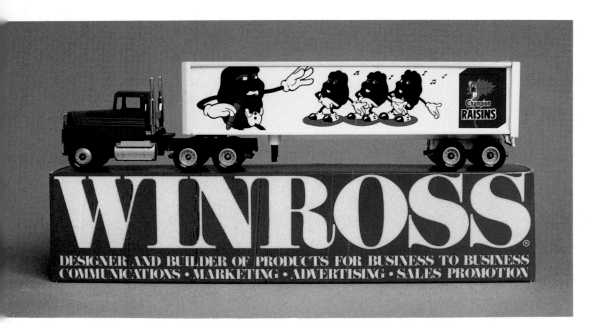

Eighteen-wheel Tractor-trailer *Champion Raisins* truck, limited production piece by *Winross*®, die-cast truck, 10" long x 2.5" high, tractor and trailer separate. The bottom of the trailer reads "The New American Highway Series," a series of replicas of actual trucks on the road. Truck sits on standard orange Winross box. Value $250 MIB.

Champion Raisins scene appears on both sides of the Winross® tractor-trailer. Box is reversed and describes the fine workmanship and detail of these vehicles.

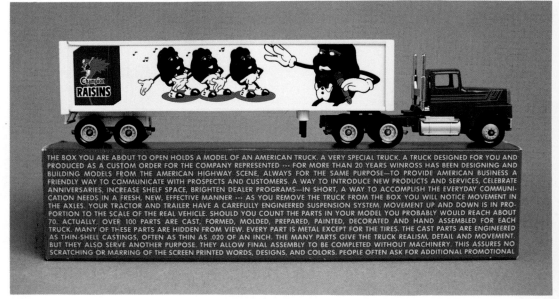

View from above.

Sun-Maid **Big Red Box** Creative Fun Kit, written in French. Box is 8.25" x 6.75" x 2.5" and has a plastic carrying handle, copyright 1987 CALRAB, Applause. To acquire this kit, consumers could mail in $5.99 and two proofs of purchase from two six-packs of Sun-Maid raisins. (*Information Courtesy of Don & April Barcus*)

Sun-Maid **Big Red Box** Creative Fun Kit contents. Kit has games and projects for children ages 4-9, including crayons, and small PVC (same as released by Hardee's). *Note:* the booklet with raisin on cover is written in English and French; this box was bought in Canada. Value $100-125.

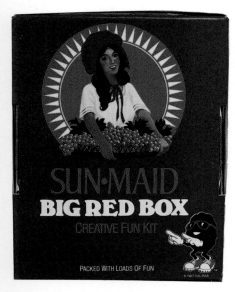

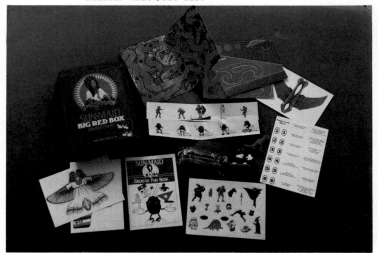

Reverse side of Big Red Box, written in English.

Sun-Maid **Big Red Box** contents. *Note:* the booklet with raisin on cover is written in English only; this box was bought in the U.S.A. Value $100-125. (*Courtesy of Toni Crittenden & Art Voorhees*)

Magic-Catch Mitts in 11.5" x 13" box with Velcro balls. Value $50-60 MIB; $20-25 loose. (*Courtesy of Barry Ullmann*)

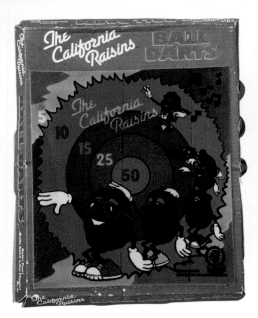

Colorforms Play Set in blue box, shown open.
Value $10-12.

Ball Darts in blue box, three Velcro balls,
18.5" x 15" x 2.5". Value $22-25. (*Courtesy of
Stan, Dayle, & Allen Golomski*)

Colorforms Play Set in purple 12" x 7.75"
box. Value $10-12.

Crayon-by-Number in 13" x 10" box. Value
$18-20.

Ball Darts in yellow box. Value $35-45 MIB.
(*Courtesy of Barry Ullmann*)

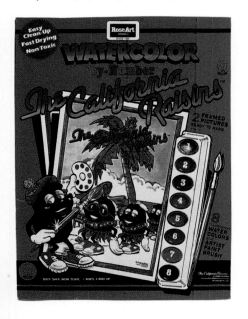

Cartoon Creator Poseable Stamper Playset
in box. Value $45-50 MIB; $35-40 used. See
contents on following page. (*Courtesy of
Barry Ullmann*)

Watercolor by-Number in 13" x 10" box.
Value $20-25. (*Courtesy of Barry Ullmann*)

Cartoon Creator box contents.

Clay Factory in 11" x 8" box. Value $40-45 MIB; $30-35 used. (*Courtesy of Barry Ullmann*)

PlayInns tent by Wenzel is 72" x 36" x 30", box 7.25" x 15", shown MIB. Value $90-110 MIB; $35-40 used. (*Courtesy of Barry Ullmann*)

Tambourine and Kazoo in 9.5" x 6" blister pack by Imperial. Value $22-25 MIP; $8 each piece used.

Board Game in 18.75" x 9.5" box with different cover, shown open. Value $15-18 used.

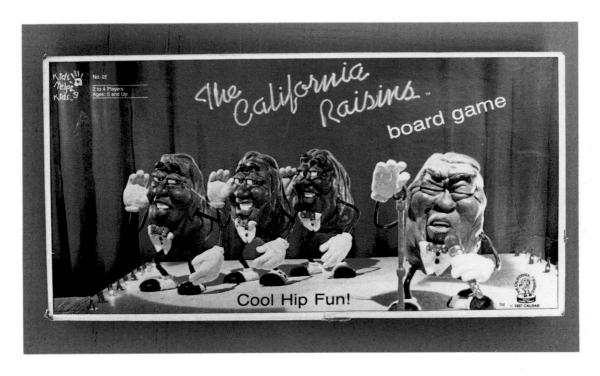

Board Game in 18.75" x 9.5" box, this was offered to fan club members for $10.95 plus $2 shipping, a savings of over $6 off the department store price. Value $15-18 used.

Playing board for the Board Game.

Chalk Board with different scene, used. Value $12-15.

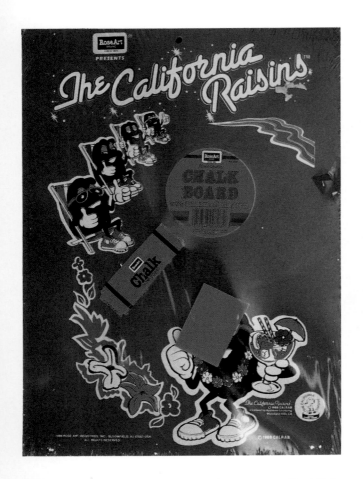

Chalk Board in package with chalk and foam eraser, 16" x 12", by Rose Art. Value $18-22 MIP

Jigsaw Puzzle, 500 pieces in 14" x 14" box, shown MIB factory sealed, American Publishing. Value $15-18 MIB factory sealed; $8-10 opened.

Jigsaw Puzzle, 125 pieces, by American. Value $30 MIB factory sealed; $12-15 opened.

Jigsaw Puzzle, 75 pieces in 4.5" x 4.5" box, shown MIB, American. Value $30 MIB factory sealed; $12-15 opened.

Shrinky Dinks in 9" x 6" purple box by Colorforms. Value $12-15 MIB.

Shrinky Dinks in 9" x 6" blue box by Colorforms. Value $12-15 MIB.

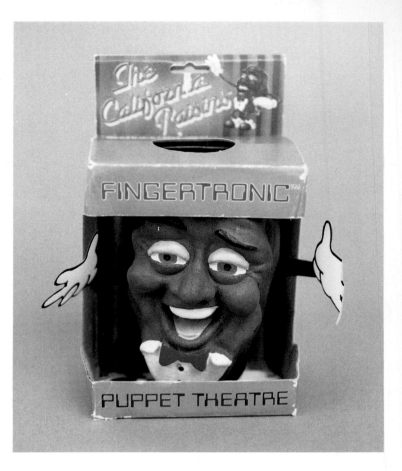

IBM Computer Game in 9.25" x 6.75" box by Box Office, ©1988 CALRAB. Value $20-25.

Fingertronic puppet in 6.5" x 5" box (cellophane missing), male raisin with purple eyelids, blue tie. Value $20-25.

Back of computer game box.

Fingertronic puppet, yellow sunglasses and shoes. Value $35-40 MIB.

120

Fingertronic puppet, orange sunglasses and shoes, $20-25 MIB; Male with "funny face," $35-40 MIB.

Female pink shoes and female yellow shoes, $30-35 each MIB.

Female green shoes, and male in tux with blue eyelids, raisin puppets out of box, but still attached to background stage from box. Puppet is foam, arms and legs are thin plastic. As shown, with arms and legs still attached, no box, value $10-12 each.

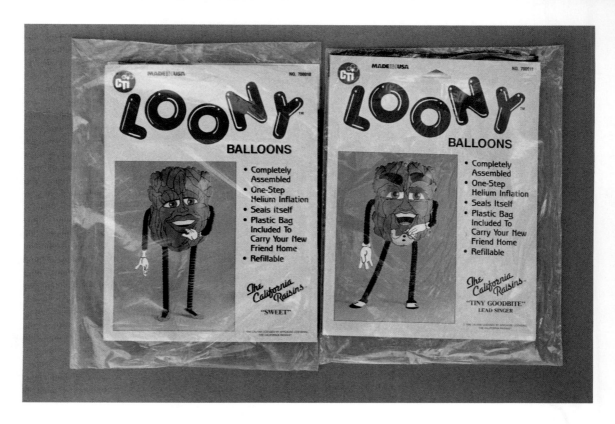

Loony Balloons of *Sweet* and *Tiny Goodbite*, shown MIP. Value $22-25 each.

Playing Cards, full deck in yellow box, was available free from *Post Raisin Bran* cereal for four proofs of purchase, or $1.95 and one proof of purchase. Value $5-7.

Inflatable Raisin, shown MIP, by Imperial. Value $30-35 MIP; $15-20 used.

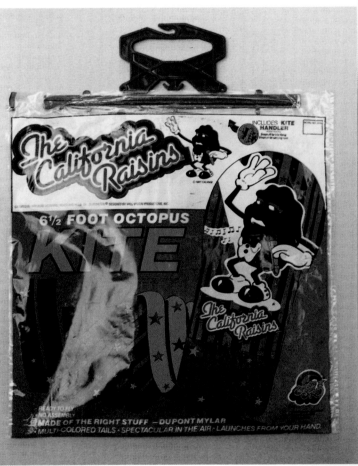

Kite in blue plastic bag with handle.

Kite shown mint in 15.5" x 16" pink plastic bag with handle. Value $20-25 MIP; $8-10 used.

Information for Collectors

In the interest of documenting what is available in the field of collecting California Raisins™, we would appreciate collectors contacting us if they have items not pictured in this book. All merchandise must be licensed by the California Raisin Advisory Board (CALRAB). Photographs and descriptions of what you have found would be welcome.

We would also be interested in documenting the various unlicensed raisin products that appeared on the market. Pictures of any of these that you may have would be welcomed. If you are unsure of the status of an item you own, send us a clear photo and description, and we'll try to make a determination for you. Along with your letter and photo, we request that you submit a self-addressed, stamped envelope (S.A.S.E.) for our reply.

Two California Raisin™ collectors, Barry Ullmann of Florida and Don Barcus of North Carolina, are interested in forming a collectors club and publishing a newsletter. If you are excited about sharing information, trading or selling extra items, and becoming part of a collectors' network, then write to us. All correspondence must include a S.A.S.E. for club and newsletter information to be sent to you. Until an official club can be formed and an address established, send all correspondence to the authors and it will be forwarded.

Pam and George Curran
P. O. Box 713
New Smyrna Beach, FL 32170-0713
Phone (904) 760-6600
Fax (904) 760-5004

Fan Club Merchandise for Members

Fan Club T-shirts, 1988, available to members only, in adult S, M, L, XL sizes only, for $8.50, which included shipping and handling.

1988 Holiday Collector's Mugs, offered only during the 1988 holiday season while supplies lasted. Set of four festive mugs featuring different scenes, in box, for $15.85, which included shipping and handling.

California Raisin AM/FM Radio, 1988, shaped like a raisin, in box, for $19.95, which included shipping and handling.

California Raisin Board Game, 1988, four singers on stage, for $10.95 plus $2 shipping and handling. This was a special price for fan club members, less than the $17.50 department store price.

California Raisin Beach Towel, 1989, 30" x 60", beach scene of "Surf's Up," for $10, which included shipping and handling.

California Raisin Cap, 1989, adjustable size in off-white corduroy with a colorful California Raisin on front, for $5.50, which included shipping and handling.

California Raisin 1990 Calendar, 1989, 11½" x 11½", cover features the Hip Band and the girls, Sweet, Delicious, and Marvelous, for $7, which included shipping and handling.

Raisin Pillow People, 1989, 26-inch long pillow, for $10.50, which included shipping and handling.

Lady California Raisin Rug, 1989, moon-shaped rug 29" x 17", pictures the girl raisins waitressing in a diner, for $7.50, which included shipping and handling.

Ms. Marvelous, 1989, PVC figurine, 3" high, wears green shoes and bracelet, has a tambourine held down to floor position in right hand, for $1.50, which included shipping and handling.

Bibliography

Applause® Licensing, Inc., printed literature and catalogs.

California Raisin Advisory Board printed literature.

The California Raisins™ Fan Club printed literature.

The California Raisins' Grapevine Gazette, Volume 1, No. 1, 2, 3, 4, 5, 6.

The California Raisins™ *Meet The Raisins*, copyright Will Vinton Productions, Claymation®.

The California Raisins™ *Raisins: Sold Out!* copyright Will Vinton Productions, Claymation®.

Nabisco Foods Inc. printed literature.

Post® Natural Raisin Bran, General Foods, Inc., printed literature.

Seto, Benjamin, *The Fresno Bee*, copyright February 21, 1995.

Sun-Maid® Raisins printed literature.

Photos from your collection

Photos from your collection